WE BOTH READ®

Parent's Introduction

Whether your child is a beginning, reluctant, or eager reader, this book offers a fun and easy way to support your child in reading.

Developed with reading education specialists, We Both Read books invite you and your child to take turns reading aloud. You read the left-hand pages of the book, and your child reads the right-hand pages—which have been written at one of six early reading levels. The result is a wonderful new reading experience and faster reading development!

What is this word?

Can you try to sound it out?

This is a special bilingual edition of a We Both Read book. On each page the text is in two languages. This offers the opportunity for you and your child to read in either language. It also offers the opportunity to learn new words in another language.

In some books, a few challenging words are introduced in the parent's text with **bold** lettering. Pointing out and discussing these words can help to build your child's reading vocabulary. If your child is a beginning reader, it may be helpful to run a finger under the text as each of you reads. To help show whose turn it is, a blue dot ● comes before text for you to read, and a red star ★ comes before text for your child to read.

If your child struggles with a word, you can encourage "sounding it out," but not all words can be sounded out. Your child might pick up clues about a difficult word from other words in the sentence or a picture on the page. If your child struggles with a word for more than five seconds, it is usually best to simply say the word.

As you read together, praise your child's efforts and keep the reading fun. Simply sharing the enjoyment of reading together will increase your child's skills and help to start your child on a lifetime of reading enjoyment!

Introducción a los padres

No importa que su hijo sea un lector principiante, reacio o ansioso, este libro ofrece una manera fácil y divertida de ayudarlo en la lectura.

Desarrollados con especialistas en educación de lectura, los libros We Both Read invitan a usted y a su hijo a turnarse para leer en voz alta. Usted lee las páginas de la izquierda del libro y su hijo lee las páginas de la derecha, que se han escrito en uno de seis primeros niveles de lectura. ¡El resultado es una nueva y maravillosa experiencia de lectura y un desarrollo más rápido de la misma!

¿Qué palabra es?

¿Puedes intentar pronunciarlo?

Esta es una edición especial bilingüe de un libro de We Both Read. En cada página el texto aparece en dos idiomas. Esto le ofrece la oportunidad de que usted y su hijo lean en cualquiera de los dos idiomas. También le ofrece la oportunidad de aprender nuevas palabras en otro idioma.

En algunos libros, se presentan en el texto de los padres algunas palabras difíciles con letras **en negrita**. Señalar y discutir estas palabras puede ayudar a desarrollar el vocabulario de lectura de su hijo. Si su hijo es un lector principiante, puede ser útil deslizar un dedo debajo del texto a medida que cada uno de ustedes lea. Para mostrar de quién es el turno para leer, encontrará un punto azul ● antes del texto para usted, y una estrella roja ★ antes del texto para el niño.

Si su hijo tiene dificultad con una palabra, puede animarlo a "pronunciarla", pero no todas las palabras se pueden pronunciar fácilmente. Su hijo puede obtener pistas sobre una palabra difícil a partir de otras palabras en la oración o de una imagen en la página. Si su hijo tiene dificultades con una palabra durante más de cinco segundos, por lo general es mejor decir simplemente la palabra.

Mientras leen juntos, elogie los esfuerzos de su hijo y mantenga la diversión de la lectura. ¡El simple hecho de compartir el placer de leer juntos aumentará las destrezas de su hijo y lo ayudará a que disfrute de la lectura para toda la vida!

Trucks • *Camiones*

A Bilingual We Both Read® Book in English and Spanish
Level K–1
Guided Reading: Level C

English text copyright © 2024 by Sindy McKay
Spanish text copyright © 2025 by Treasure Bay, Inc.
Use of photographs provided by iStock and Dreamstime
We Both Read® is a trademark of Treasure Bay, Inc.

Published by
Treasure Bay, Inc.
PO Box 519
Roseville, CA 95661 USA

Printed in China

Library of Congress Catalog Card Number: 2024939967

ISBN: 978-1-60115-073-8

Visit us online at
WeBothRead.com/Bilingual

PR-10-24-4.5

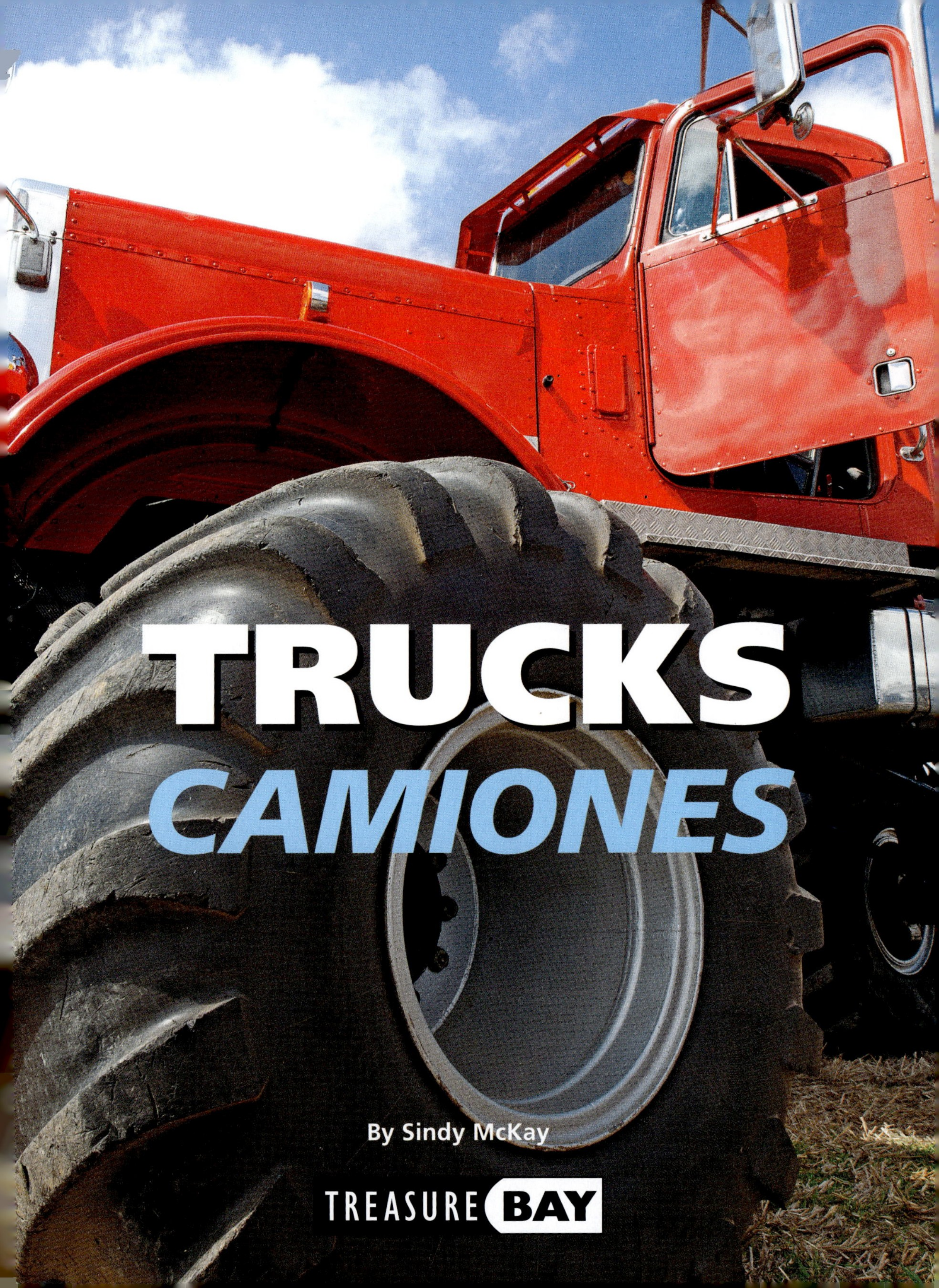

TRUCKS
CAMIONES

By Sindy McKay

TREASURE BAY

● Honk! Honk! It seems like everywhere you go there are trucks on the road!
You see pick-up trucks, delivery trucks…

———————◆———————

¡Tuuu! ¡Tuuu! Parece que hay camiones rodando donde quiera que vayas. Puedes ver camionetas, camiones de reparto…

★ . . . and big rigs!

———————◆———————

. . . ¡y tráileres!

Trucks are large vehicles that can carry things from one place to another.

Maybe you have spotted one that is carrying food. At a special loading dock, food is put...

———————◆———————

Los camiones son vehículos muy grandes que pueden cargar cosas de un lugar a otro.

Tal vez hayas visto alguno cargando comida. En un muelle de carga especial, se sube la comida...

⭐ . . . in the truck.

———————◆———————

. . . *al camión.*

● The truck then drives to the **store** where it may back into a receiving dock. Here the food is unloaded and goes into…

———————— ◆ ————————

*Luego, el camión se dirige a la **tienda** donde entra en retroceso al muelle de recepción. Aquí, se descarga la comida para llevarla dentro de…*

★ ...the **store**.

———————◆———————

... *la* *tienda*.

They can carry logs, or space shuttles, . . .

———————◆———————

Pueden cargar troncos, o naves espaciales . . .

★ . . . or cars.

———————— ◆ ————————

. . . o carros.

A truck can help you **move** to a new **house**.

———————◆———————

*Un camión puede ayudarte a **mudar** a una nueva **casa**.*

★ Or **move** the **house** to you!

◆

*¡O **mudar** la **casa** hacia ti!*

11

Tanker trucks **haul** liquids in big tanks. The tanks may contain water, milk, or gas.

———————◆———————

*Los camiones cisterna **transportan** líquidos en tanques grandes. Los tanques pueden contener agua, leche o gasolina.*

★ Dump trucks can **haul** rocks.

◆

Los camiones de volteo a veces ***transportan*** *piedras.*

Most mail is delivered in a mail truck. Mail is delivered in all kinds of weather, even when there's . . .

——————— ◆ ———————

La mayoría del correo se entrega desde una camioneta de correos. El correo se reparte en todo tipo de clima, hasta cuando hay . . .

★ . . . snow.

— ◆ —

. . . *nieve.*

These trucks deliver purchased items directly to your home. Some of the items are made in far off lands then shipped across the ocean in containers on…

---◆---

Estos camiones entregan los productos que compras directo a tu casa. Algunos de los productos se fabrican en tierras lejanas y luego cruzan el océano en contenedores dentro de…

★   . . . big ships.

———————◆———————

. . . *buques grandes.*

When the ships arrive at the port, the containers are put onto large container trucks to be taken to warehouses.

From the warehouses, these items are carried by truck to various stores. Now they are ready...

———————◆———————

Cuando los buques llegan al puerto, los contenedores se colocan en camiones grandes que los llevan a los almacenes.

De los almacenes, estos productos se llevan en camiones a varias tiendas. Ahora están listos...

★ ...to sell!

———————◆———————

...¡para vender!

Many trucks bring things to you. Some, like garbage and recycling trucks, take things away.

This truck takes crushed aluminum cans away to the recycling center.

———————◆———————

Hay muchos camiones que entregan cosas. Algunos, como los camiones que recolectan la basura y los reciclables, se llevan cosas.

Este camión lleva latas de aluminio aplastadas al centro de reciclaje.

★   This truck picks up trash bins.

———————◆———————

Este camión recolecta los contenedores de basura.

Your garbage isn't the only thing trucks remove. They can remove snow.

They can remove dead branches. And, if you park in the wrong place, they can remove . . .

———————◆———————

Tu *basura no es lo único que se llevan los camiones. Pueden quitar la nieve. Pueden llevarse ramas secas. Y, si te estacionas en el lugar equivocado, pueden llevarse . . .*

NO PARKİNG

★ ...**your** car!

...*¡tu carro!*

23

• Certain trucks help keep us safe. Hook and ladder fire trucks are used to rescue people from tall buildings and other high places. Other fire trucks carry hoses that help put out fires by spraying lots of...

---◆---

Algunos camiones nos mantienen seguros. Los camiones de bomberos con escalera ayudan a rescatar a personas de los edificios y lugares altos. Otros camiones de bomberos cargan mangueras que ayudan a apagar incendios rociando mucha...

★ . . .water.

. . .*agua.*

An armored truck is used to safely transport money. It is like a giant metal safe.

An ambulance, or emergency services truck, takes people to get medical care.

———————◆———————

Una manera segura de transportar el dinero es con un camión blindado. Es como una caja fuerte gigante hecha de metal.

La ambulancia, o camión de emergencias, transporta a las personas para que reciban atención médica.

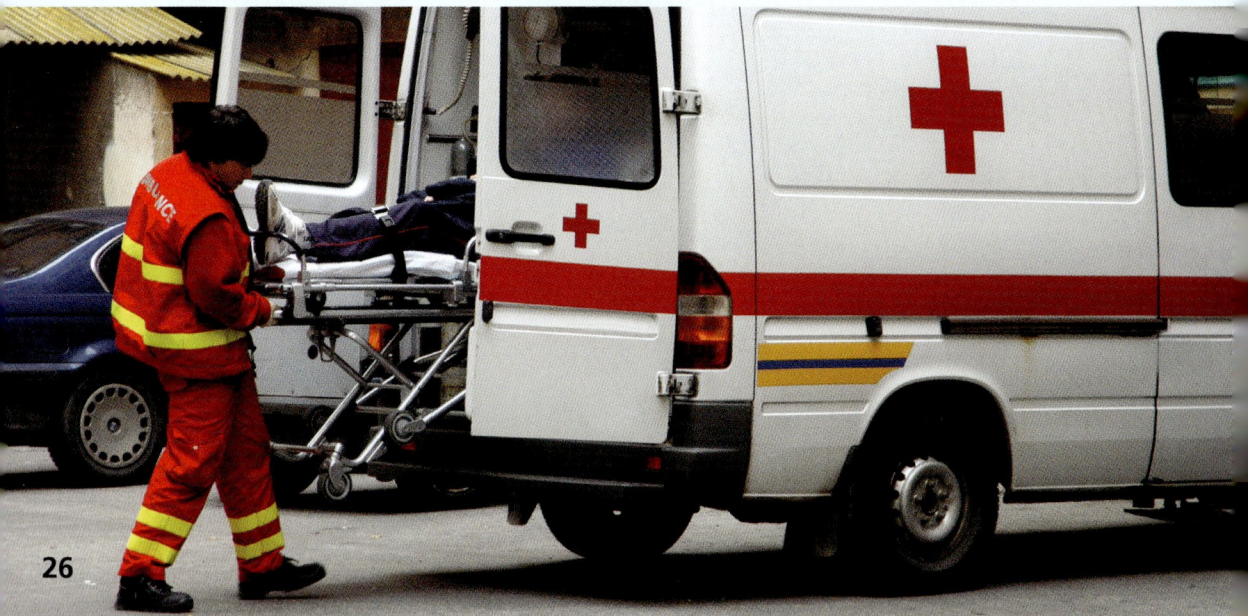

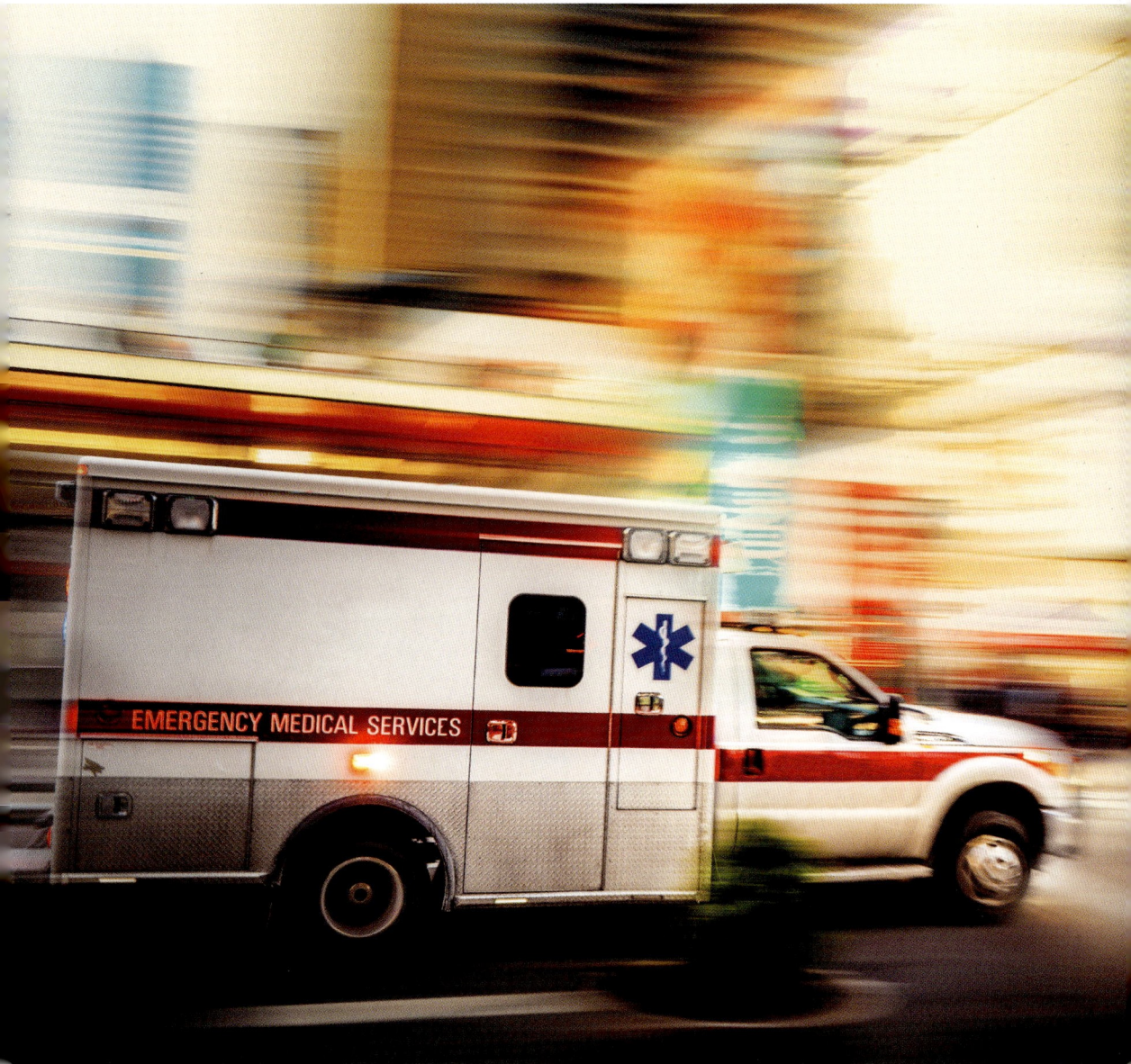

★   It can go fast!

◆

¡Puede ir rápido!

Workers use boom bucket trucks to reach **high** places to do repairs, wash windows, trim trees, and many other important jobs.

———————◆———————

*Con un camión de plataforma elevadora, los trabajadores pueden llegar muy **alto** para hacer reparaciones, limpiar ventanas, recortar árboles y hacer otros trabajos importantes.*

★ That is **high**!

♦

*¡Qué **alto**!*

Some trucks can lift their containers high in the air to load supplies onto jet planes.

Some have a long, low **flatbed**. Very large items can be carried...

———————————— ◆ ————————————

Algunos camiones pueden alzar sus contenedores en el aire para llevar suministros a los aviones de propulsión.

*Otros tienen una **plataforma** larga y baja. ¡Se pueden cargar cosas muy grandes...*

⭐ …on a **flatbed** truck!

◆

…*en un camión de **plataforma**!*

Trucks are especially useful on a farm. This hay baler rolls the hay into cylinders which are then loaded onto a flatbed behind a pick-up truck.

———————— ◆ ————————

Los camiones son particularmente útiles en una granja. Esta embaladora enrolla la paja en cilindros, los cuales se cargan en una plataforma detrás de una camioneta.

★ It takes the hay away.

─────────◆─────────

Se lleva la paja.

Food trucks can be found just about anywhere that people get hungry.
Taco trucks are especially popular!

———————— ◆ ————————

Puedes encontrar camiones de comida donde pueda haber gente con ganas de comer.

¡Los camiones taqueros son bastante populares!

★   This truck sells ice cream.

◆

Este camión vende helado.

● **Monster** trucks are specially made pickups with **cool**-looking oversized tires that allow them to race over rough terrain. They often do amazing stunts where they drive over old cars, crushing them.

———————◆———————

*Los camiones **monstruo** son camionetas especiales cuyas enormes y **geniales** llantas permiten que corran sobre terrenos escabrosos.*

★ **Monster** trucks are **cool**!

◆

*¡Los camiones **monstruo** son **geniales**!*

There are many types of trucks and many ways to use them. The next time you are on the road, see how many different kinds you can see.

———————— ◆ ————————

Hay muchos tipos de camiones y se utilizan de muchas maneras. La próxima vez que vayas por la calle, a ver cuántos tipos diferentes logras encontrar.

⭐ **Honk! Honk!**

¡Tuuu! ¡Tuuu!

Glossary • *Glosario*

big rig • *tráiler*
a truck, sometimes called a semi or a tractor-trailer, which has 18 wheels and a large trailer

un tipo de camión, también conocido como semirremolque o camión de carga, que tiene 18 llantas y un contenedor grande

tanker • *camión cisterna*
a vehicle with a large tank used to transport liquids, such as milk or gas

un vehículo con un tanque grande que se usa para transportar líquidos como leche o gasolina

armored truck • *camión blindado*
a vehicle used to transport cash or other valuable items

un vehículo que se usa para transportar dinero en efectivo u otras cosas de valor

boom bucket • *plataforma elevadora*
a bucket-shaped device attached to a truck with a long extension arm that is used to reach high places

una plataforma en forma de cubeta que se sujeta a una camioneta con un brazo de extensión para poder alcanzar lugares altos

flatbed • *camión de plataforma*
a truck with a low, flat area for carrying large, long items

un camión con un área baja y plana para transportar cargas grandes y largas

monster truck • *camión monstruo*
a specialized off-road performance vehicle with four-wheel steering and huge tires

un vehículo todo terreno especializado y de alto rendimiento que tiene dirección en las cuatro llantas y llantas enormes

Questions • *Preguntas*

Add to the benefits of reading this book by discussing answers to these questions. Also consider discussing a few of your own questions.

Aumente los beneficios de leer este libro al conversar sobre las respuestas de las siguientes preguntas. También podría formular sus propias preguntas y discutirlas.

1 What are some differences between trucks and cars?
¿Qué diferencias hay entre camiones y carros?

2 Name as many things as you can that a truck might carry.
¿Qué podría cargar un camión? Nombra tantas cosas como puedas.

3 If you were to drive a truck, what kind of truck would you most like to drive? Why?
Si pudieras manejar un camión, ¿cuál es el que más te gustaría manejar? ¿Por qué?

4 What kinds of equipment might you find on a fire truck?
¿Qué tipo de equipos podrías encontrar en un camión de bomberos?

If you liked *Trucks,* here are some other
We Both Read® books you are sure to enjoy!

Si les gustó **Camiones**, *¡seguramente disfrutarán
de estos otros libros de* We Both Read®*!*

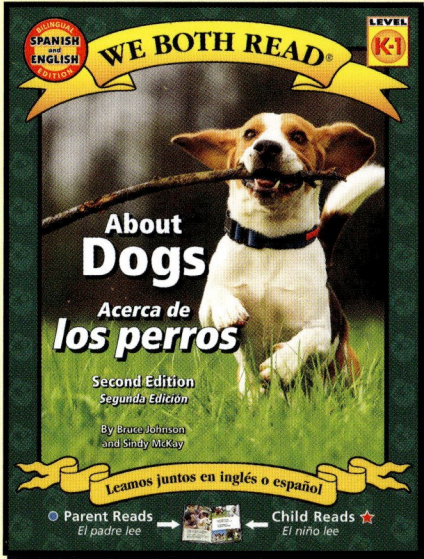

BILINGUAL
SPANISH
and ENGLISH
EDITION

WE BOTH READ®

LEVEL
K·1

About
Dogs
Acerca de
los perros

Second Edition
Segunda Edición

By Bruce Johnson
and Sindy McKay

Leamos juntos en inglés o español

● Parent Reads → Child Reads ★
El padre lee   *El niño lee*

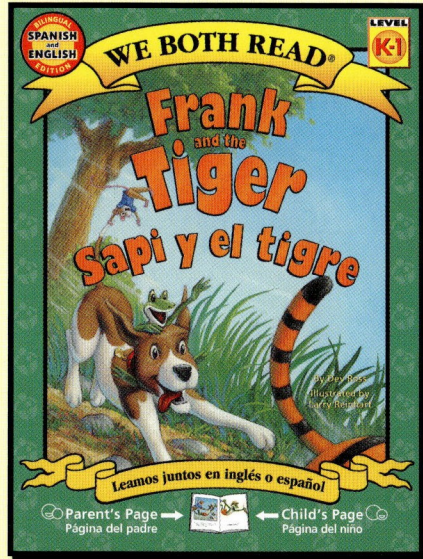

BILINGUAL
SPANISH
and ENGLISH
EDITION

WE BOTH READ®

LEVEL
K·1

Frank
and the
Tiger
Sapi y el tigre

By Dev Ross
Illustrated by
Larry Reinhart

Leamos juntos en inglés o español

Parent's Page → Child's Page
Página del padre   *Página del niño*

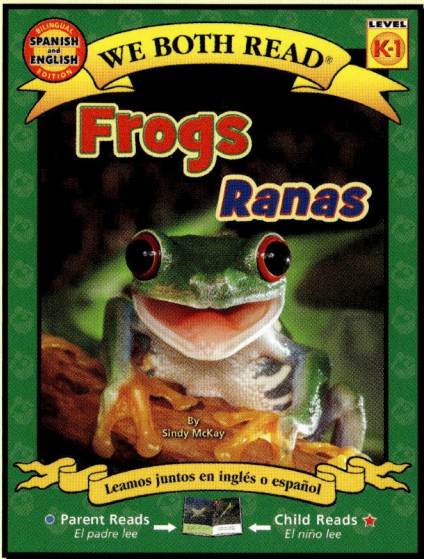

BILINGUAL
SPANISH
and ENGLISH
EDITION

WE BOTH READ®

LEVEL
K·1

Frogs
Ranas

By
Sindy McKay

Leamos juntos en inglés o español

● Parent Reads → Child Reads ★
El padre lee   *El niño lee*

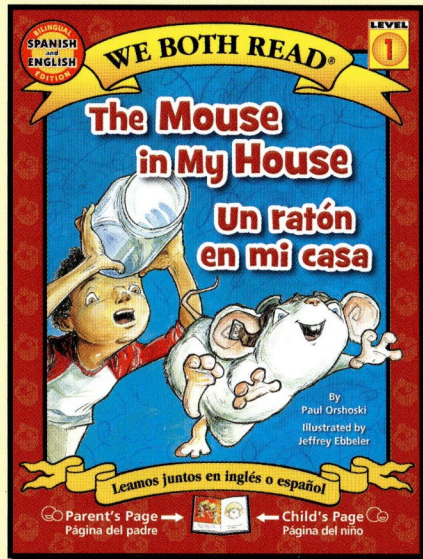

BILINGUAL
SPANISH
and ENGLISH
EDITION

WE BOTH READ®

LEVEL
1

**The Mouse
in My House**
**Un ratón
en mi casa**

By
Paul Orshoski
Illustrated by
Jeffrey Ebbeler

Leamos juntos en inglés o español

Parent's Page → Child's Page
Página del padre   *Página del niño*

You can see all the We Both Read books
that are available at **WeBothRead.com**.

*Visita el siguiente sitio web para descubrir todos
los libros disponibles de **WeBothRead.com**.*